SECRETS OF APPLIANCE REPAIRMEN

A CONSUMER'S HANDBOOK ON SAVING MONEY

by
RAYMOND O. WOODBURY

Published by
RAYMOND O. WOODBURY

With special thanks to my wife, JoAnn, for her patience and understanding during the preparation and writing of this manuscript

Library of Congress Control Number: 2002091266

ISBN 0-9718947-0-1

First Printing:
200 Copies April 2002

Copyright ©2001
Raymond O. Woodbury

All Rights Reserved. No part of this publication may be reproduced, stored or transmitted in any form or by any means, electronic, mechanical photocopying, recording or otherwise without the prior written permission of the copyright owner

Printed in the U.S.A. by
Morris Publishing
3212 E. Hwy. 30
Kearney, NE 68847

PREFACE

What is this book about? It's about saving money - your money! This book is the most money-saving book that you will ever read. The suggestions are authoritative and to the point. You will learn the simple secrets that service technicians employ every day, plus there are many examples of preventive maintenance that you can do yourself. If, after you have read this book, you save yourself the price of just one service call, this book will have paid for itself. This book will explain the simple procedures you can do on your own, to correct problems with your appliances. Every appliance in your home will be covered in this book. Use the index to find your problem, and then follow the directions on what to do to correct it. Each appliance has it's own little set of problems that prevent it from functioning properly. You will learn the secrets that the service technician uses every day. Every problem covered in this book is what the service technician calls "easy calls" or "nuisance calls".

If the simple corrections listed in this book do not help, you may have to call a service technician, as your problem may be more serious than this book covers. This book is NOT A REPAIR MANUAL. If you must call for service, this book will cover your relationship with the service person, and how to get the most for your money. There is a list of dos and don'ts when dealing with the service person.

Many of the most common problems are the result of faulty installation, which may not be recognized until after the warranty has expired. Other problems develop from a lack of normal maintenance on certain appliances. Before you call for service, search through the index first. You may be able to solve your own problem. Most appliances today however, are technologically advanced and are hard to understand for repair purposes. Even the directions for how to use them are sometimes confusing. A well-trained technician can explain the ins and outs of using and programming most appliances. If you don't understand how to do the programming, then call for a qualified technician before the warranty expires.

If you decide to replace the appliance and not make repairs, then ask friends and relatives what brand they own, and how they like it's performance, before you shop for a new one. When you are in the store, ask the sales person to show you how to operate it, what all of the controls are for, are the written operating instructions easy to follow. Also ask about the warranty. What is covered and for how long? Take as much time as you need before you buy. Make sure you understand how to operate it, and ask yourself if you need the extra controls, or gadgets. Make sure you will be satisfied with your new appliance before you make your purchase. That is, make sure it meets your needs.

CAUTION: Price alone should not be the major criteria when purchasing your appliance. Ask whether the store services the merchandise it sells. Always try to go to a dealer that has a good reputation, and one that provides parts and service. If you must have your appliance repaired, call a service company with a good reputation. Chances are you will get good, honest, and guaranteed service. When in doubt about the reputation of any company, call your local Better Business Bureau first.

When you do call for service, keep in mind that the person that you want to enter your home is a professional, who takes pride in their work, so try to clean the work area surrounding the appliance and to allow plenty of room. Remember, your home is their work place just as your office or factory is your work place. First impressions are lasting impressions, so try to make their first impression of you and your home a positive one.

If you have the time, it is recommended that you read the entire contents of this book, in order to get an idea of how to avoid future problems with your appliances. Helpful hints are contained in every chapter and may help you understand how some appliances work. I sincerely hope this book helps you in some way.

TABLE OF CONTENTS

CHAPTER   PAGE

1. **DO NOT REPAIR** 1
2. **A LIST OF DO'S AND DON'TS** 3
3. **DISHWASHERS** 5
   Dishes Not Clean; Water Temperature; Not Enough Water; Not Pumping Water Out; Odors; Leaking; Foaming; Dishes Not Dry.
4. **GAS RANGE** 15
   Burners Won't Light; Oven Won't Come On; Won't Bake Properly; Baking Hints.
5. **ELECTRIC RANGE** 20
   Oven Cleaning; Oven Won't Heat; Checking the Thermostat.
6. **REFRIGERATOR** 23
   Preventive Maintenance; Not Cooling; Sweating Around Doors; Ice Formation; Freezer Compartment; Odors.
7. **TRASH COMPACTORS** 31
   Proper Use Of; Proper Loading.

8. **GARBAGE DISPOSER** 33
   Noisy; Unit Jammed; Won't Grind; Water Won't Drain; Grease in Drain.

9. **FREEZER** 39
   Chest Type Freezer; Not Freezing Properly; Freezer Warm; Upright Freezer; Chart for Freezing Foods.

10. **AIR CONDITIONER** 50
    Window Units; Not Cooling; Central Air Conditioners.

11. **FURNACES** 55
    No Heat; Caution; Oil Heat; Filters; Changing Belts.

12. **AUTOMATIC WASHER** 61
    Won't Pump Out; Washer Shuts Off; Wet Clothes; Water Keeps Running; Water Enters Very Slow; Other Problems; Helpful Hints; Detergents - Use Of; Odors.

13. **DRYERS** 70
    Slow Drying; Won't Start; Odor on Clothes; Combination Washer/Dryer;

14. **FUSES** 76
   Finding the Fuse; Replacing Fuses; Circuit Breakers; Extension Cords; Screw in Type Fuses; Amperage Rating.
15. **HUMIDIFIERS** 82
   Why We Need Them; How Do They Work; Cleaning.
16. **DEHUMIDIFIERS** 86
   No Water From Drain; Helpful Hints and Caution.
17. **HOT WATER TANKS** 89
   How to Flush the Tank; Running Out of Hot Water; Hints on Replacing the Old Tank.
18. **MISCELLANEOUS** 93
   Fluorescent Lights; Electric Can Openers; Toasters; Cabaret Fans; Garage Door Opener.
19. **A GLOSSARY OF TERMS USED** 103

Chapter 1 — DO NOT REPAIR

In addition to your major appliances, you have many small appliances in your home that are subject to breakdowns. Most of them are impossible to repair without the knowledge of how to disassemble them, and where to get repair parts for them. Some have plastic cases that are snapped together at the factory, and to figure out how they come apart takes a certain skill. These include coffee makers, blenders, mixers, vacuum cleaners, toasters and many other small appliances in the home. It is not recommended that repairs be attempted on small appliances, as the appliance could be ruined and of no further use to you. Check the phone book for "small appliance repair" shops and give them a call first. Get an estimate for the repair and decide if it should be repaired or replaced. The cost of repairs can sometimes exceed the price of a new appliance. In some states the electric utility company will repair certain electrical appliances at no charge, but you pay for parts used. It's worth a phone call to find out if your local company will make repairs for you.

CAUTION: Never attempt repairs on micro-wave ovens, television, V.C.R., Stereo or other electronic device. Most of them have high voltage involved and can be dangerous, even if the plug has been pulled from the wall. In addition to the high voltage most are very sensitive and can be rendered inoperable with the slightest mistake.

LARGE APPLIANCES:

More and more of the large appliances have electronic circuit boards in place of mechanical timers and should not removed by anyone but a qualified service technician. They can be so sensitive that a static spark from the finger can ruin them.

Most appliances will give many years of satisfactory service if they are not abused. Keep them clean, do not overload and don't try to use them beyond what they were designed for. An appliance is a lot like a car; that is, the better they are maintained the more miles you can expect of them.

Chapter 2: A HELPFUL LIST OF DO'S AND DON'TS

DO'S:

1. Do stay home the day service is scheduled.
2. Do answer the door promptly.
3. Do inform the service person if you are handicapped and will take longer to answer the door. They may think no one is at home.
4. Do make arrangements in advance for payment with the service company.
5. Do keep the pets and children safely away from the work area.
6. Do make sure an adult is home at the time of service. Most reputable service persons will not enter our home without an adult present.
7. Do provide notice to the service company at least one day prior to the scheduled day if you must cancel or can't be home.
8. Do demand the old or exchanged parts, unless the parts must be returned when under warranty.

DON'TS:

1. Don't keep the service person waiting at the door for a long period of time. Their time is valuable. It could be raining or freezing cold outside also.

2. Don't complain to them. If you are unhappy with the service company, call the service manager. Don't take it out on the service technician. If the technician is at fault, call the service manager.

3. Don't be a grouchy boss. Don't try to repair it yourself. You could make matters worse.

4. Don't try to help. They are the experts.

5. Don't expect them to enter your home when no one is present. They won't, so don't leave a note on the door asking them to come in, make repairs and send a bill.

6. Don't ask them for favors. Their time on the job determines the price of repairs. If they falsify their time they could loose their job.

Chapter 3 DISHWASHERS

Most problems with dishwashers can be taken care of by you with a little logic and common sense. The most common complaints are simple in nature and once you understand the cause, the cure is easy.

DISHES NOT CLEAN:

The most common complaint is, dishes not getting clean. First, look at your detergent. Is it fresh? Is it hard and lumpy? Powder dishwasher detergent will lose its strength if it is left open and is more than thirty days old. Then, it has a tendency not to dissolve properly and leave a film on the dishes. Also, make sure the spray arms are turning freely and are not obstructed in any way.

WATER TEMPERATURE:

Another frequent problem is the water temperature. It should be maintained at 140 to 160 degrees at all times. If you cannot hold your hand in the water, it's just right. Dishwasher detergent will not dissolve in cool water; therefore will not clean the dishes. In most cases the water line to the dishwasher is installed to the hot water line going to the hot water faucet on your sink. Portable models are hooked up directly to the faucet. In either case, run the water from the faucet until the water is about as hot as it will get before starting the dishwasher. This practice will ensure that the first fill cycle will have plenty of hot water, which is very important for clean dishes.

NOT ENOUGH WATER:

Check your Owner's Manual for the right amount of water that should be in the bottom after filling. If you do not have the recommended amount then look for a restriction in the waterline. Is the valve opened all the way? The valve should be located under the sink on the HOT water line. If the valve is open all the way and you have good pressure at the sink faucet, you could have a clogged screen in the inlet valve to the dishwasher, located under the dishwasher. (Figure 1, P. 12). The copper water line could also have a kink. You need a qualified person to correct it.

NOT PUMPING THE WATER OUT:

This can be caused by several reasons. First check the drain hose for a kink. Look under the counter, beneath the sink or under the dishwasher itself. The drain hose should not have any sharp bends in it. After a year or more the hose can collapse at the sharp bend and restrict the water flow. If this is what you find, then you can correct it yourself, if you can replace the hose. Another possibility is the part that is called an air gap (Figure 2, P. 12). It is located on the sink or counter top. When the dishwasher empties, you should hear water rushing through it. If not, take the cover off by grasping it and pulling up with a slight twisting motion. Inside you will find there is a cap that will unscrew counter clockwise. On some models you must pry the cap off. After you have removed it, you will find a smaller tube inside the larger one. Clean the debris out of the small tube and the cap, re-assemble, and try the dishwasher again. It should pump out okay. If it still will not pump out, then go to the user manual and follow directions on how to clean the pump screen (if your model has one). Does your dishwasher drain to a garbage disposer? Have you replaced the disposer recently? If so, did you remember to remove the knockout plug before putting the dishwasher drain hose on? If you didn't it cannot drain. The last but most important consideration is the pump. Did it make a very loud noise before it stopped pumping? A foreign object, such as a metal screw or other hard material, could have stripped the pump impeller. If that is the problem, then you need a qualified service technician to repair it for you.

When shopping for a new dishwasher, you will find that the different models are so varied that it will be hard to make up your mind as to what you really want. You have to decide if you want the one with the computer or a manual control. Be sure you can understand how to use it before you purchase it. If you don't understand it, you will dislike like it when you get it home.

BAD ODOR:

If you have a built-in dishwasher, then check the drain hose very carefully. If it is fastened to the disposal and goes downward and to the dishwasher you could be pumping dirty water into the dishwasher every time you run the disposal. The hose should be fastened to the bottom side of the counter top to prevent this. All dishwashers do not have a check valve to stop a back flow. Check valves are an internal part of the pump housing so it can't be seen by the naked eye. An air gap also prevents back flow. Install an air gap to be sure.

PORTABLE DISHWASHERS:

Portable dishwashers have the same problems as the built-in models except the drain hose hooks directly to the faucet. If the water drains very slowly, the hose may be kinked inside the cabinet. Pull the hose out as far as it will go, to eliminate any kinks. The spray arms must me clean, free of debris, and spin freely when pushed by hand. This rule includes portable and built-in models.

LEAKING ON FLOOR:

Most leaks can be diagnosed and solved by you if you know the simple rules. There are several reasons for the water to spill out of the dishwasher. One is the safety float assembly. This is the round (usually white) object in the bottom of the tub. (Figure 3, P. 13). It should be located in either the right front, or the left front corner by the door. The float is what the name implies that is; it must float on top of the water. It controls a switch on the underside of the dishwasher that, in turn, controls the amount of water entering on each fill cycle. If the float is sticking in the low position, too much water will enter and cause an overflow. If it sticks in the high position, no water will enter. Check the float first for a leaking problem. Some floats will lift out so they can be cleaned. Simply lift up, but do not force it. It may be fastened on the underside of the dishwasher to prevent it from being lost. If one or more spray arms are not turning, the water spray can be directed toward the door seal, or toward the bottom of the door. If this happens, it can cause seepage, or a flood. Always check the spray arms, (Figure 4, P.13) and the dishes that may come in contact with them. The spray arms must move freely.

FOAMING:

Another cause for leaks is foaming. When foaming occurs, there will be a large amount of suds inside the dishwasher. Foaming is the result of the wrong detergent being used either by accident, or from not following directions on the proper amount to use. Use only automatic dishwasher detergent in powder or liquid form. When the proper detergent is used, there should be **NO** suds of any kind during the wash cycle.

Never use dish soap. If you use the wrong kind of detergent and produce lots of suds, you can kill the suds very easy. Pour a cup of cooking oil into the bottom of the dishwasher, close the door, and restart the motor. Let it run for two or three minutes, then pump the water out by turning the timer knob, or pushing the "cancel drain" button. Repeat if necessary. If all the above suggestions fail, you could have a bad door seal, a leaking pump seal or other problems. Call for a qualified service technician.

DISHES NOT DRY:

Most dishwashers have a heater in the bottom of the tub. The heater will come on during the dry cycle, unless there is a control to shut it off. The control is usually marked "Hot Dry" or "Cool Dry". When the heater is "on", it heats the air to a temperature suitable to dry the moisture in the air. To dry properly, the water must also be hot enough to aid in the drying process. All the moisture will not dry up in any dishwasher. The secret is wait until the dry cycle is just about finished, open the door about 4 inches, and let the dishes air dry. You can do this before the dry cycle starts also. Remember, hot dishes will dry by themselves, if hot enough when the door is opened. Try it both ways to see which is best for you-that is, after the dry cycle or before it starts. (Figure 5, P.14)

Figure 1 - Inlet Valve Screen

Figure 2 - Air Gap With Cover Off

Dishwashers 13

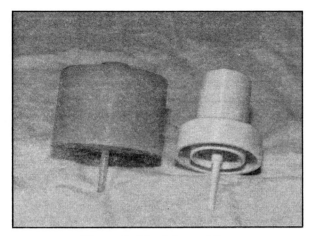

Figure 3 - Float Assemblies

Figure 4 - Spray Arm

Chapter 4   GAS RANGE

The gas range is another appliance that has become technologically advanced in recent years. Many of the controls are electronic and hard to diagnose. It takes a lot of training to properly repair many of the new models. Most of them must be plugged into a 120-volt outlet to operate.

BURNERS WON'T LIGHT:

If you have the electronic ignition and it does not spark when you turn the knob, then check the outlet to be sure you have power to the range. Plug a lamp or radio in to see if it works. Power OKAY? When you cleaned your range the last time did you remove the knobs and the cams under the knobs? (Figure 6, P.18) If you did, then check to see if you put the cams back on the wrong way. If you did, the igniter can't work. Remember how they go back on so you will not have the problem in the future. If you can hear the spark when the knob is turned but the burner won't ignite, you most likely have a pilot hole plugged up. (Figure 7, P. 19) Remove the burner and look

Secrets of Appliance Repairmen 16

real closely on the side towards the spark unit and look for a small hole or orifice. If it is plugged up, then clean it very carefully. Do not use a toothpick because it could break off inside the hole, or orifice. Instead, use a straight pin or needle. Wash it out with soap and water, dry it completely, then re-install the burner and try again. If you have a standing pilot and the burner won't ignite, clean the burner as described above.

OVEN WON'T COME ON:

First check the outlet (see above) first. Got power? Check the clock setting and follow directions for manual operation. It should come on. If not you may need service, as an igniter or sensor may be bad.

WON'T BAKE PROPERLY:

Most oven thermostats are very accurate and last for years, but occasionally they may need adjustment. But before this is done you can try something else first. Most hardware stores and appliance stores sell thermometers that you place on the baking rack. Set the thermostat for 350 degrees and let the oven heat for approximately 20 minutes, and then check the

Gas Range 17

temperature. If it is within ten degrees of 350 then it is okay. You may have to adjust your recipe or baking utensil.

BAKING HINTS:

A dark pan or a glass dish will bake faster because they will absorb more heat. An aluminum or light colored pan will take longer to bake because they reflect heat. It is best to adjust **baking time, not the temperature.**

If you have a self-cleaning oven and it will not work, check the instructions on the console, or on the clock. If they are not clear, then read the operating manual very carefully. Some models have two or three steps that must be taken before the self-clean mode can be obtained.

It is easy to forget how to set properly, so don't feel bad if you have to re-read the instructions every time. After all, you do not do this every day.

CAUTION: If you have the "continuous clean" model, with the grainy surface, never use oven cleaner on this surface. Instead, follow directions, or turn the oven on "high" for two hours and let the oven clean itself. More than one time may be needed.

If you must call for service, then have the stovetop clean

Secrets of Appliance Repairmen

and everything out of the oven. Give the service technician plenty of room to work and have it reasonably clean. If you do not have a self-cleaning oven and must clean it manually, with oven cleaning products made for that purpose, read the directions very carefully. If the directions say to have plenty of ventilation, then please do so. Some products have ingredients that can be harmful if you breath too much of the fumes. You can become unconscious in a matter of seconds. These products are not dangerous if used as directed by the manufacturer.

Figure 6 - Ignition Cam

Gas Range 19

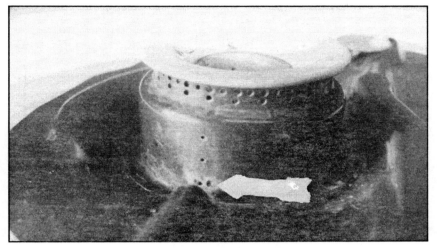

Figure 7 - Pilot Orifice

Chapter 5 ELECTRIC RANGES

Electric ranges are very complicated and hard to diagnose as well as repair. They are best left to the service technician who is qualified to work on them. There are only a couple of things that you can do for yourself. If you find that the burners and the oven won't come on, but the clock and the lights work then you may have a bad fuse. Refer to Chapter 14, Page 76, on fuses for help. If you are having problems with baking, then refer to the chapter on gas ranges.

We have all heard the stories of how Great Grandmother used to bake on an old wood burning stove, and things came out perfect every time. Well, for the most part this is true. She did not have a thermostat to set, nor did she have a thermometer to put in the oven to tell her when the temperature was just right. She knew what baking dish or pan to use, depending on what she was going to bake. She also kept a wary eye on what she was baking and took it out of the oven at just the right time. So if you are having a problem, follow Grandmother's example and you can have perfect results too.

Electric Range

Of course, if your oven won't heat up properly, your thermostat is likely at fault. You can't bake in a barely warm oven. Call for service to have it repaired, or adjusted.

CAUTION: Never take a panel off to have a look; you could get a nasty shock. Also never try to repair a broken or burned wire. There is a right way and a wrong way to do that, so leave it to the professional who knows what to do and how to do it.

CLEANING THE OVEN:

See "caution" in Gas Range Chapter 4, Page 16.

OVEN WON'T HEAT:

First, check the clock for the proper setting. Set it for "manual", turn the selector switch for "bake" and set the thermostat for 350°F. The oven should come on. If it still won't heat, try the clock setting again. An alternate check would be to set the control for "broil" instead of bake. If the broiler comes on, the bake element could be burned out. If "bake" or "broil" do not work, the control or thermostat may be at fault. Call for professional help.

If the clock and oven light are the only thing working on

Secrets of Appliance Repairmen

the range, you could have a blown fuse in your fuse box. An electric range requires two fuses to operate the heating elements, but the lights and clock will usually continue to operate with only one fuse. See Fuse Chapter 14, Page 76.

CHECKING THE THERMOSTAT:

You can check the thermostat for accuracy if you have a candy or cooking thermometer. If you have an oven safe glass tumbler or cup, fill it 3/4 full of water and set it as close to the center of the oven as you can.

Turn the oven on and set the thermostat for 212 degrees. The control knob will not have that setting on it so make a guess at 210. Insert the thermostat into the water and wait for the oven to get hot. Water boils at 212 degrees, so if the water comes to a boil turn the thermostat down, a little at a time until the water stops boiling. Now check the thermometer in the cup. It should read from 200 to 210 degrees. If the reading is 15 degrees or more too low, the thermostat needs replacing or adjusted. The same is true if you must set the control to 220 degrees or higher to make the water boil.

Chapter 6   REFRIGERATOR

Refrigerators are getting very high-tech along with everything else these days and require qualified technicians to repair them. Some models have automatic defrost, in the door dispensers and even computers to monitor temperature. Very high-tech stuff but very difficult to repair. However, there are some things that most people can do for themselves.

PREVENTIVE MAINTENANCE:

Even if you have a very simple refrigerator without all the extras on it, you should clean the condenser occasionally. The condenser is located at the bottom, inside the grill. You should feel air blowing out of the grill. Remove the grill and, using a flashlight or a plug in light, look inside for the condenser coils. They collect a lot of dirt and lint. That cuts the airflow through the coils and if completely plugged up, the refrigerator can't get cold. Go to an appliance store or hardware store and buy a brush that is made for this job. It is about 30 inches long and about 2 inches in diameter with a tapered end. (Figure 8, P. 29). Use the brush to loosen the dirt, and then use your vacuum cleaner to remove it. Use the crevice tool; it works best. Do this once a month if it is in a high traffic area, otherwise every two months is sufficient.

The ideal way to clean the condenser is to use compressed air and blow the dirt out, but it's messy and everybody does not have compressed air at his or her disposal. So use the method described above, it's the easiest and cleanest way.

NOT COOLING:

If the refrigerator begins to warm up for no apparent reason, first check the airflow, and do the above procedure if needed. If it doesn't help, then check the airflow inside the refrigerator. Is the fan running? The air must circulate through the fan and over the evaporator (Figure 9, P. 30). A package or plastic bag can block the intake or exhaust. Don't overload with food to the point that air circulation is cut off. Also check the thermostat. Did it accidentally get set to a higher temperature? If none of the above solves your problem, then the problem may be a mechanical failure that you cannot repair. The failure could be as simple as a failed Defrost timer, or a defective thermostat. If you hear a loud clicking noise every two or three minutes coming from the area of the compressor, you probably have a failed compressor. That will cost from $300.00 to $500.00 to repair. You may want to purchase a new refrigerator, instead of making repairs. Unfortunately, someone who is qualified and understands what to do and how to do it must diagnose most problems. When you do call for service, make sure the area is clean, and there is enough room to pull the unit away from the wall, if necessary. If you have carpeting, then try to provide a piece of cardboard or something to slide the refrigerator on, so the carpet won't be damaged. The technician does not want the responsibility of damage and will appreciate your concern.

Also be prepared to empty the shelves, if required by the technician, if they must work inside the compartment.

SWEATING AROUND DOORS:

Sweating around the doors is most common in the summer when high humidity is present. This is caused when the warm, humid air comes in contact with the cold surface of the refrigerator exterior. Condensation forms on these cold areas.

Secrets of Appliance Repairmen 26

Most refrigerators have a mullion heater (See Glossary P. 103) to compensate for this and are automatic. Other models have a switch provided so it can be turned off in the dry season. The switch may simply say "off" or "on". Others will say "dry-wet", "high-low" or other words to that effect. The switch is merely an off/on switch and only has two positions. Setting the switch to the proper position usually takes care of the sweating.

ICE FORMATION:

Every time the refrigerator defrosts, it melts the frost accumulation on the evaporator plate, or coil. The water flows down through a tube to a pan under, or near the compressor. The heat from the compressor evaporates it, so it won't accumulate. If the drain tube is obstructed, the water will stay in the tray under the evaporator coil and re-freeze.

You can check the tube yourself without much trouble. First you must find it. It may be located on the rear of the refrigerator, running down into the compressor compartment.

Remove it and clean with hot water. It may also be located under the refrigerator near the condenser coils. This one is called a "J" tube because it resembles the letter J. You can remove it and clean it easily by pulling it out of the rubber grommet. You may have to twist it slightly. While you have either drain off, check the tube it came out of for obstructions. Use a plastic straw or tooth pick. You may find food particles, paper, or other debris caught inside. If everything is clean and open, the pan heater may be defective. The pan is where the water drips during the defrost cycle. The heater prevents the water from freezing in the pan. If all the above suggestions do not help, call for service before you go any further.

FREEZER COMPARTMENT:

If you have a freezer compartment it should be kept at the coldest temperature possible. The most recommended temperature is at "0" degrees. Ice cream should be brick hard. If you own a separate freezer also, then use the refrigerator freezer for short-term items only. These are described in the freezer chapter.

ODORS:

Bad odors in the freezer can be eliminated with a thorough washing of the interior. Ice cubes pick up odors very easily and should be thrown away often, or when they start to taste bad. Packages brought home from the store can also be a source of odors. Frozen food packages can pick up odors from other packages in the grocery bag. If they come in contact with a box of soap, for instance, the odor will transfer to the package, but the odors will not penetrate to the food, so that should not be a concern.

Look for torn packages where the food may be exposed to the air in the freezer. That is another source of odors.

The freezer compartment should be emptied at least once a year and thoroughly cleaned. Clean more often if necessary and don't forget to clean the ice cube tray at the same time.

Many people keep an open box of baking soda in the freezer compartment, and the refrigerator compartment at the same time, to absorb odors. Some appliance stores carry a liquid form of odor control.

REFRIGERATOR ODORS:

With the higher temperatures in the refrigerator compartment, bacteria can grow in spilled foods such as milk, soups and other liquids, as well as solid foods.

Refrigerator

Bacteria will create odors faster in the refrigerator than they will in the freezer. This means wiping up spills at once, and more frequent washing of the interior. Odors from spoiled foods will migrate to other foods very quickly, so remove any spoiled foods at once.

When cleaning the interior, do not forget to wash the shelves in the door, and equally important, wash the seal around the door also.

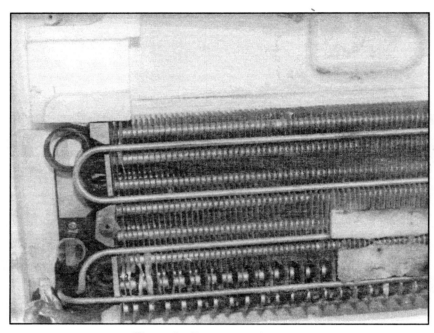

Figure 9 - Evaporator Plate

Chapter 7   TRASH COMPACTORS

Compactors are made to be so powerful that they could destroy themselves if loaded improperly. Because of this power, a series of switches is placed strategically in the unit to prevent this from happening.

This makes it almost impossible for any person to service a compactor himself or herself. It takes a well-qualified technician to diagnose and repair. So the best course of action is preventing problems from happening.

LOADING:

To ensure the proper operation be sure you load the drawer evenly. That is, don't put hard-to-crush items too far to the front, or to the rear of the drawer. If the load is uneven the drawer could be pushed out towards the front. When that happens one or more of the switches can open, shutting the unit off, so it does not destroy itself. Therefore, exercise caution when loading. A little common sense goes a long way. Don't overload to the point that the bin, or bag is spilling out over the top.

Remove the bag before it gets that full and start with a clean one. Debris spilling over the top can get into the gears or drive chain and jam the unit.

If you must call for help, then make sure nothing is on top of the compactor because the cover will have to be removed. If it is built in under the counter, then pull it out so it can be serviced. The service person does not want the responsibility of damaging the counter top or the floor.

If you have small children in the house, it's a good practice to remove the key from the start switch. Store it in a convenient place where you won't lose it, so children can't put the family cat inside to see what happens. Remember, SAFETY FIRST!

Chapter 8   GARBAGE DISPOSER

A garbage disposer unit is basically trouble free when used according to directions. However, you can have a couple of problems that can be taken care of yourself.

NOISY:

If it starts making an awful noise shut it off immediately. Using a flashlight, look **VERY** carefully into the grinding chamber, for an object that was put in by mistake, such as a piece of wire, a coin or anything that won't grind. If you find something, then remove it before turning the unit back on. Use a flashlight and long nose pliers, not your hand.

UNIT JAMMED:

If it jams and will not turn, the overload switch will kick out. (Figure 10, P. 37) First, find the wrench that was furnished with the unit, (Figure 12, P. 38) and insert it into the hole in the very bottom center of the unit. (Figure 11, P. 37) Turn the wrench in both directions, alternating left and right.

CAUTION: If you turn with too much force the whole unit could fall out of the mounting bracket. You may want to have another person turn the crank, while you check the grinding chamber for the object causing the problem. Once it is freed up, remove the foreign object before turning it on. If you look very carefully, you will find the reset button on the bottom of the unit, close to the place you inserted the wrench.

Push the button in and see if it stays in. If it does then you are ready to start using it again. If it will not start, you may have blown a fuse, or tripped a circuit breaker. See Chapter on Fuses, Page 76.

WON'T GRIND:

If the blades (providing your unit has them) are sticking and don't move freely, the unit won't grind anything. To free them up, simply feed a tray of ice cubes into the chamber while it is running. Do this as often as necessary. If the rotor turns easily and there are no obstructions in the grinding chamber, and the fuse blows, or the circuit breaker trips, you have a shorted motor. Repairs are sometimes more expensive than a new unit. You should consider purchasing a new one. Before installing a new disposer, you should first check the drain hose from the dishwasher (if you have one). If it drains into the disposer, you will find a knockout plug in the new unit. It must be removed before the drain hose is attached. If you do not remove the plug, the dishwasher will not drain. Every time you use the disposer, remember to run lots of cold water to keep the drain open. If you find you must call for service, then be sure to clean out everything from under the sink, so the service person can get to the unit. Also make sure the sink is empty.

WATER WON'T DRAIN:

If the water backs up into the sink when you try to grind something, then the drain is most likely at fault. Have the drain cleaned or replaced. When using the disposer, always remember to run plenty of water, even after you shut the unit off. This helps keep the drain open and running free.

GREASE:

If you pour cooking oil or grease into the disposer, always run lots of cold water. The grease will form into very small particles and flushes down the drain. On the other hand, if hot water is used, it will liquefy and pass through the trap under the sink, only to solidify somewhere else in the drain. If that happens, you will need a plumber to have it cleaned.

CAUTION: Never pour drain cleaners into the garbage disposer to open a drain. The seals can be damaged and the metal pitted as a result. The disposer could be ruined. Some drain cleaners contain toxic chemicals and are strong enough to damage some types of plumbing, and a garbage disposal as well. Never leave any waste in the garbage disposal for any length of time. Grind it when it is put in and use **lots of water**.

If you have a problem drain that is always plugged up, all the drain cleaners for sale will not help you. Call for a plumber and save yourself a lot of grief.

Garbage Disposer

Figure 10 - Location Of Overload Switch

Figure 11 - Location For Wrench

Chapter 9   **FREEZER**

Freezers share common problems with refrigerators. Both have thermostats, circulating fans, evaporators and condensers.

Most have defrost timers that turn on the defrost heater once every 24 hours. The big difference being, the operating temperature. There are different types of freezers.

CHEST TYPE FREEZER:

This type of freezer has the door mounted on the top, so you look down into the compartment when the door is open. These are virtually trouble free, if defrosted when needed. Most models do not have a defrost timer, and must be defrosted manually.

NOT FREEZING PROPERLY:

Too much frost build-up on the side of the walls will insulate the food packages from the cooling surface. Any insulating barrier will affect the temperature of the food. If you defrost, and the frost builds up rapidly again, the door seal may be at fault. If it is loose, warm air will migrate into the compartment, causing almost instant frost. Check for cracks or tears in the seal. Also feel for hardness of the seal. A hard, dried out seal needs replacing. If all of the above checks are negative, then do the dollar bill test. Using a fairly new bill (one with no wrinkles), place it on the surface between the seal and the top surface. Close the door on the bill and gently pull. If it pulls out with a lot of resistance, the seal is good. If you can move the bill back and forth, the door is out of adjustment. Try this test all the way around the door to get an idea of which end is loose, and which end is snug. If you find that the right end is very loose, and the left is tight, you can adjust the door in the following manner. With the door open, place your hands on both ends of the door and pull down with your right hand while pushing up with your left. Not too hard now, you could break the plastic lining. Go easy as it may take more than one try. Each time you do the adjustment try the dollar bill again until the bill fits the same all the way around the door. If the left end is loose, reverse the process. You can also put a flashlight inside with the room lights off and look for light around the gasket.

FREEZER WARM:

If the freezer starts to warm suddenly, check for power at the plug. You can do this by plugging in a floor lamp in the same outlet. If you have power to the freezer, and it is warming up, you may have a control failure. Call for a qualified technician at once. If the food is still cold and has not started to thaw, you can save it by purchasing some dry-ice. Look in the yellow pages under Ice supply.

CAUTION: DO NOT TOUCH DRY ICE WITH YOUR BARE HANDS: It is super cold and can freeze the skin immediately. Always use insulated gloves or tongs to handle it.

UPRIGHT FREEZER:

Most models defrost every 24 hours, with the defrost water going to a pan under the compressor, where it evaporates from the heat of the compressor. Be sure the drain is clean, as explained in the chapter on refrigerators. Some upright freezers have the condenser under the box, while others have the condenser on the back of the box. In either case, the condenser must be kept clean at all times for proper cooling. If the condenser is located on the rear, it will extend almost from the top to the bottom of the freezer. There is no fan on this type of condenser. It relies on the air in the room to cool it. If the airflow is interrupted, the gas inside the condenser will not cool. Check for paper bags, newspapers or any other material that may have fallen on top of, or inside of, the space where the air flows.

Also, make sure the freezer is not pushed against the wall. This could further restrict the airflow. This type of condenser needs air flow on both sides to work properly

FREEZING FOODS

TEMPERATURE:

Your freezer should maintain a temperature of 0° F or below. Maximum temperature should be no more than 5° F.

Check temperature with freezer thermometer or outdoor thermometer, or use this rule of thumb; if freezer can't keep ice cream brick solid, the temperature is above the recommended level. In this case, don't store food more than a week.

TIME:

Date food packages with an "expiration date" according to maximum storage time recommended below. Longer storage is not dangerous, but flavors and textures begin to deteriorate.

PACKAGING:

Use foil, moisture vapor-proof plastic bags and wraps, freezer wrap or freezer containers. Foil, when folded, may develop pinholes, resulting in freezer burn.

COMMERCIAL FROZEN FOODS:

Pick up frozen foods just before going to the check-out counter. Purchase only foods frozen solid. Place in your home freezer as soon as possible. Cook or thaw according to the label instructions.

HOME FROZEN FOODS:

Freeze in coldest part of freezer, at the bottom. Freeze no more than 3 lbs. per cubic foot of freezer space within 24 hours. Don't freeze a quarter of beef all at once, as explained above.

STORAGE CHART

Bacon	*
Corned beef	*
Frankfurters	+
Ground beef, lamb, veal	2-3 months
Ground pork	1-2 months
Ham and picnic, cured	*
Luncheon meat	+

ROASTS

Beef	12 months
Lamb, veal	6-9 months
Pork	3-6 months
Sausage, dry, smoked	+_
Sausage, fresh, unsalted	1-2 months
Steaks and chops, beef	6-9 months
Lamb, veal	3-4 months
Pork	2-3 months
Venison, game birds	8-12 months

* Freezing cured meats not recommended. Saltiness encourages rancidity. If frozen, use within a month.

+ Freezing not recommended. Emulsion may be broken and

product will weep, that is, the oils and moisture will seep out.

+_ Freezing alters flavor.

Fish-- home frozen and purchased frozen

Fillets and steaks from "lean" fish; cod, flounder, haddock sole	6 months
"Fatty" fish; bluefish, perch, mackerel salmon	2-3 months
Breaded fish	3 months
Clams	3 months
Cooked fish or seafood	3 months
King crab	10 month
Lobster tails	3 months
Oysters	4 months
Scallops	3 months
Shrimp, uncooked	12 months

Keep purchased frozen fish in original wrapping; thaw; follow cooking directions on label.

Poultry-home frozen or purchased frozen chicken, whole or cut up .... 10 months

Chicken livers .... 3 months

Cooked poultry .... 3 months

Duck, turkey .... 6 months

FRUITS AND VEGETABLES

Fruit-home frozen or purchased frozen berries, cherries, peaches, pears, pineapple, etc. .... 12 months

Citrus fruit and juice frozen at home .... 6 months

Fruit juice concentrates .... 12 months

Freeze above in moisture vapor-proof container.

VEGETABLES-HOME FROZEN OR PURCHASED FROZEN

Home frozen .... 10 months

Purchased frozen- cartons, plastic bags or boil-in-bag ....
.... 8 months

Cabbage, celery, salad greens and tomatoes do not freeze successfully.

COMMERCIAL FROZEN FOODS

Also see: meats, fish poultry, fruits, and vegetables.

Baked goods

Yeast bread and rolls baked.................................3-6 months

Rolls, partially baked...2-3 months

Bread, unbaked...1 month

Quick bread, baked...2-3 months

Cake, baked, unfrosted Angel food .... 2 months

Chiffon, sponge .... 2 months

Cheesecake...2-3 months

Chocolate... 4 months

Fruitcake .... 12 months

Yellow or pound .... 6 months

Cake, baked, frosted ....8-12 months

Cookies, baked ....8-12 months

Pie, baked ....1-2 months

Fruit pie, unbaked .... 8 months

Freezing does not freshen baked goods. It can only maintain the quality (freshness) the food had before freezing.

Main dishes

Meat, fish, poultry pies and casseroles .... 3 months

TV dinners, shrimp, ham, pork, frankfurter .... 3 months

Beef, turkey, chicken, fish .... 6 months

HOME FROZEN FOODS

Also see: meats, fish, poultry, fruits, and vegetables.

Bread .... 3 months

Cake .... 3 months

Casseroles-meat, fish, poultry .... 3 months

Cookies, baked and dough .... 3 months

Nuts salted .... 6-8 months

Unsalted .... 9-12 months

Pies, unbaked fruit .... 8 months

FROZEN FOOD KNOW-HOW

Thawing-- It's best to thaw frozen fish, poultry or meat in the refrigerator. That way the surface does not reach dangerously high bacteria levels before the product thaws in the center.

Another benefit of slower thawing is less moisture (drip). Meat, fish and poultry can be cooked without thawing; allow about one third to one-half more cooking time.

Refreezing- Most partially thawed foods refreeze safely if they still contain ice crystals and are firm in the center. However, many foods (partially thawed ice cream) will not be top quality. Meat, fish and poultry purposely thawed in the refrigerator and kept no more than one day may be refrozen. Don't refreeze thawed meat or poultry pies or casseroles, cream pies or vegetables.

Food completely thawed (intentionally or by accident) and warmed to room temperature (72°F), should be thoroughly cooked immediately or discarded. Fruit and juice concentrates are exceptions; they ferment when spoiled so you can rely on taste to warn you. Toss them if flavor is "off".

** Source: MICHIGAN STATE UNIVERSITY, by permission.

Chapter 10   AIR CONDITIONER

WINDOW UNITS:

Air conditioners come in all sizes. Some will cool an entire floor in your house, and others will cool one room only. The larger units use 240 volts and require special wiring. The smaller units plug into the standard household current without a special plug. Both sizes share common problems. With all units you should follow directions in the Fuse Chapter (P. 76) if it won't start.

NOT COOLING:

If this is your problem, then do the following. Remove the cover, or grill and check the filter. If it's dirty, clean it with a mild soap solution like dish soap, and rinse it out with clear water. Before putting the filter back in, look at the evaporator (the part the filter covered). If it's dirty, clean it with a soft brush. Never use a wire brush. Follow the line of fins in the evaporator and brush in the same direction as the fins. The fins are the very thin, flat, metal plates spaced close together. If you brush across

Air Conditioner

the fins, they could bend. Most evaporators look similar to a car radiator. However, some look like a twisted metal brush, with little pieces of aluminum pointing in every direction. Never use a brush on this type of evaporator. Instead, make sure it is dry, and then use a vacuum cleaner with a soft brush on the end to clean it. When you are done cleaning, return the filter and cover and try again to see if it will cool. Still won't cool? Go outside and see if you can reach the air conditioner, then feel for air blowing through the condenser. (Figure 13, P. 54) You should have a fairly, strong and even airflow over the entire condenser. It should be blowing hot air. If you have very little airflow, you have found the problem. Shut the unit off and pull the plug from the wall.

Using a household cleaner that has a trigger type sprayer, like Top Job, adjust it for direct spray, not a mist. Soak the condenser with the cleaner. Be sure you spray between the fins. Let set for about ½ hour and repeat, letting it set about 15 minutes more. At that time take the garden hose and squirt it directly into the condenser with as much force as possible, to wash out the loosened dirt. **DO NOT SPRAY INTO THE SIDE OPENINGS.** This is only a temporary repair. Sooner or later it will have to be taken apart and cleaned properly.

Let it dry out for an hour, plug in and try again. It should cool. If not, call for an air conditioning specialist and have it diagnosed properly.

CENTRAL AIR CONDITIONERS:

Central air conditioners share the same type of problems; blown fuse, dirty filters and dirty condenser units. Always check the furnace filters (Figure 14, P. 55) and replace them when dirty. Outside, check the condenser for leaves and grass on the condenser and clean if needed.

Air conditioners rarely need to have Freon added, (Glossary Page 103) but if everything is running and clean this may be the problem.

As always, check the controls and be sure they are set properly before you call for help. On central air conditioners, check the thermostat setting first. It should be set for "cool", and the fan control set for "automatic". Also, the switch on the side of the furnace must be turned on. Some communities require and additional shut off switch on the outside of the house, near the compressor unit (Figure 13, P. 54). Be sure it is turned on. Most

Air Conditioner

condenser fan motors (the unit outside the house) are controlled by a thermostat. When the weather is very hot, the thermostat will turn the fan motor on to a higher speed to compensate for the higher temperatures. If the thermostat fails, the air conditioner will run, but it will not cool properly. If you think the fan is not running fast enough, you must have someone who is qualified to diagnose and repair it. If the fan is running fast enough but still not cooling, check for airflow through the condenser. The air should be pulled in through the sides and out the top. If there is no air coming out the top, the chances are the condenser is plugged up with dirt and dust and must be cleaned.

If the compressor will not start, check the fuses as described in the chapter on fuses. Page 76.

Remember, air conditioners are the same as refrigerators; the airflow must be sufficient over the condenser and the evaporator to cool properly.

Secrets of Appliance Repairmen 54

Figure 13 - Condenser Location

Figure 14 - Furnace Filter

Chapter 11   FURNACES

Natural gas furnaces used to be very simple, and some models still are, but today with the high efficiency gas models being sold, you can't do much except regular maintenance on them, cleaning and changing filters regularly. The high efficiency types have blowers, switches and safety devices on them, and are very technical to diagnose and repair.

NO HEAT:

If your furnace is the kind with no pilot light and you experience a no heat condition, all you can do is shut the switch off (on the side of the furnace), for about a minute and then turn it back on. Be sure the thermostat is turned up higher than the room temperature. The control is designed to automatically re-set, and re-light. If it will not re-light, or you do not hear a clicking noise, check for a blown fuse or tripped circuit breaker first. If the fuse is good, do not do anything. Shut the switch off and call for service. If you have a constantly burning pilot then follow directions on the furnace for re-lighting the pilot. If the pilot will not stay on, the thermocouple could be bad and should be replaced.

CAUTION: Liquid propane, or L.P. gas furnaces are the same as natural gas, except for the fuel used. If you have an L.P. gas furnace then **NEVER, NEVER TRY TO LIGHT THE PILOT.** The reason for this is that if the pilot should go out and the safety device in the valve fails, the gas will continue to flow out of the pilot orifice. The safety shuts off gas flow to the pilot if the flame goes out, but they have been known to fail. Liquid propane is heavier than air and will pool on the floor, creating a dangerous

Furnaces 57

situation. Shut the gas line valve off and call for a qualified furnace technician at once.

OIL HEAT:

Check the fuse, replace if necessary and do nothing else. Oil can be explosive if conditions are right, so try no adjustments, and do not try to re-start the furnace. Call for expert help. Our heating systems are the most important appliance in our house, but they must be treated with the utmost respect.

FILTERS:

Keep clean filters in at all times for efficiency and for the best airflow. Filters are located in the blower compartment and come in a variety of sizes. Replace with the same size that you remove and do not try to substitute with the wrong size. When purchasing filters you must decide if you want the throw away type or the reusable type. The reusable kind can be washed and used again and again. When installing filters look for the arrow on the end or side of the filter. It indicates the airflow through the filter. Replace or clean filters at least once a month. While you are changing filters, it's a good idea to oil the motor and blower at the same time. Look at both ends of the motor for a small opening with a cap on it. Put four or five drops of good grade oil in each opening. The blower also has an opening at each end of the shaft, but may not have a cap on it. Use the same oil as above at the beginning of the heating season. Use a zoom spout oiler for hard to reach places. (Figure 15, P. 59).

CHANGING BELTS:

If your furnace uses a belt to drive the blower, check the tension on the belt by moving the belt up and down to see how much play there is. It should not exceed 1½". If it does, replace it with the exact replacement required. If you look on the flat side of the belt a number should be visible. Use this number to purchase the new belt. An official substitution number is acceptable but save the box for future reference. If you can't read the number on the belt and you do not have the book that came with the furnace, you can still replace it. Shut the furnace off with the switch located on the side of the furnace (Figure 16, P. 60), remove the belt and take it with you to a furnace dealer or repair shop so they can measure it for replacement.

CAUTION: The rule of thumb is, if you do not understand it, do

Furnaces

not try to repair it. Leave it to the trained technician to repair it safely.

Trained technicians can diagnose problems, and make proper adjustments, and do repairs properly according to the condition encountered. Heating units are too dangerous to trust yourself to repair. Follow the above advice on keeping filters changed, and the area clean around the furnace.

Secrets of Appliance Repairmen 60

Figure 16 - Shut Off Switch

Chapter 12 AUTOMATIC WASHER

Automatic washers are getting very advanced and are difficult to diagnose, that's where this book will be of help. Many of the new washers have electronic controls that set water temperature, type of wash to be done, second rinse and more. However they all share the same problems that can be solved by you. They will be covered one at a time.

WON'T PUMP OUT:

First check for a kinked drain hose. The pump can't force the water through a kink so it will stay in the tub. If the motor doesn't start you may have a bad lid switch or other mechanical problem. If the motor runs and the drain is okay you could have a jammed pump or broken belt. You will need help for this type of repair.

WASHER SHUTS OFF:

Some water pumps out then washer **STOPS.** Some models have a cool-down during the permanent press cycle.

Before the wash cycle is done, it will pump out about half the water, and then pause for a minute or two, then start filling with cold water. This is an automatic function and is normal.

WET CLOTHES:

Water in the bottom of the washer when it shuts off; check the drain in your sink. Does it drain very slowly? If it does, then look at the washer drain hose. If it extends to the bottom of the sink, the water can siphon through the hose back to the washer. (Figures 17 & 18, P. 68 & 69) You will have to clean the drain or raise the end of the drain hose, so the end of it is not in the water.

WATER KEEPS RUNNING:

Again, check your drain hose. The end could be too low and cause the water to siphon out of the washer. Laying it on the floor can do the same thing. If you have a standpipe, the hose could be pushed too far down inside. Without an air break the water will siphon into the house plumbing. In both cases the water is entering the washer and emptying out at the same time.

WATER ENTERS VERY SLOW:

First, check the pressure in other areas of the house to see if it's also low. If it is, then you can't help the washer. If the pressure is normal every place else then you can check out the washer very easily. First, find the faucets where the fill hoses are screwed on. Shut the water off and remove the hoses. Check the ends of the hoses for a screen. If it's dirty clean it and replace the hose on the faucet. If you find no screens, take the other end of the hoses off the valve located on the washer and look inside. There you will find screens. Use a small rubber syringe and flush the screen with water. If they do not come clean, then try a small brush. If you must remove the screens, be very careful, because if you bend them they will not go back in straight.

CAUTION, BE SURE YOU MARK THE HOSES AND RETURN THEM TO THE SAME PLACE YOU REMOVED THEM.

OTHER PROBLEMS:

If the washer will not do anything, then follow the advice in the chapter on dryers, and check for power first.

Won't Spin? If the washer will not spin there could be any number of problems such as; a broken wire, a bad solenoid, bad lid switch, bad timer, bad clutch, loose belt, or other mechanical problem will require help.

Won't Agitate? For problems with agitation it would be the same as above.

Dispensers Won't Work? If the dispensers will not work, then it could be a bad control or it maybe plugged up with soap, fabric softener or something else. Most of the above examples you cannot repair yourself. Call for a qualified repairperson and let them do it right. After you call for service, clean off the top of the washer and move everything away from the washer. Try to have a clean floor. Also, be sure to remove any water left in the washer. If it has to be moved it will not be too heavy for the service person to handle.

HELPFUL HINTS:

Are clothes tearing? Most washers will not tear clothing under normal circumstances. If you have a mechanical problem it could happen, but first look elsewhere. Most of us do not realize how old certain articles of clothing are. If something is tearing then grasp it at the tear, fingers on both sides of the tear and gently pull. If it separates easily then it is old and the washer is not to blame. Every time bleach is used it weakens the material. Repeated washings also wear the material out and weaken it.

DETERGENTS:

Most of us use too much detergent without realizing it. The purpose of detergents is to break down the surface tension of the water. A demonstration of this is to float a double edge razor blade on the surface of the water. If you want to try this, use a dish or a cup full of water. It can be done, honest. This is surface tension. But it cannot be done if you add detergent to the water first. The surface tension is gone.

So all it takes is to add enough detergent to break the surface tension so water will flow easily through the material, carrying the dirt with it. There are additives in the detergent such as whiteners, perfumes for odor, bleaches and softeners, but the detergent in the water does the actual work. So if a little is good, more is not necessarily better. Too much suds can also cause the pump to "suds lock", and stop pumping the water out of the machine.

If you think you are using too much detergent, there is a

simple test you can do to determine this. Select some whites, such as socks, T-shirts or handkerchiefs, put them in the washer and begin filling with **HOT** water. While it is filling pour in some powder water conditioner, usually about a cup (follow directions on box) and let washer run for about 10 minutes. If you get lots of suds you are definitely using too much detergent. The suds are the residues left in the clothing. If you get little or none then you are probably using the proper amount of detergent.

HINT: To help keep your drain running clean, always put a screen on the end of the drain hose. (Figure 19, P. 69) They can be purchased at most supermarkets and hardware stores. If you cannot find them, then tie an old nylon or pantyhose to the end of the hose. It works just as well. You will be amazed at how much lint was going down your drain.

ODOR INSIDE WASHER:

Stale odors emanating from the inside of the washer when the lid is raised is very upsetting. The source of the odor is in the space between the basket, and the outer tub. The likely cause of the odor is water standing in the bottom of the tub. It can get stale in a couple of days and give off a bad smell. The probable reason for the standing water is because it did not all drain. The washer may not be level so the water cannot drain to the pump. To check, lay a level on top of the washer, first across the front, then on the left and right sides. Level the washer according to which side is too low or too high. Once you have it leveled, the odor should go away on it's own as you use the washer. To help get rid of the odor you may want to pour in one or two cups of bleach, or vinegar. Set the control for low water level, fill the washer with water and add the bleach or vinegar, but do not add clothes.

Set the timer for normal cycle and let it run the entire cycle. Another reason for water lying in the bottom of the tub is, water re-circulating back into the tub during the drain cycle. Some washers use a filter mounted to the side of the tub, and can leak water back into the tub during the drain phase. Even if the washer is level, water will lie, in the bottom, all the time when not in use.

For this kind of problem you need a trained technician to repair or replace the defective parts.

In all cases, use a little common sense before you pick up the phone to call for service. You can solve most problems, if

Secrets of Appliance Repairmen 68

you think them out first. Remember, service is expensive, so don't panic when something goes wrong. **THINK FIRST.**

Figure 17 - Drain Hose Improperly Installed

Automatic Washer 69

Figure 18 - Drain Hose Installed Properly

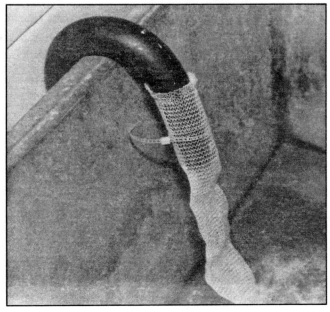

Figure 19 - Drain Hose Screen

Chapter 13 DRYERS

How does a dryer work? The drum part of the dryer tumbles the clothing so that at any given time some articles of clothing are suspended in mid-air for a moment as they fall from top to bottom in the drum. Then hot air is injected via a gas burner or heating coils. The hot air is passed through the material as it falls, releasing moisture in the form of vapor or steam. The blower then sucks the moisture-laden air out of the drum, and pushes it into the exhaust vent to the outside of the house.

SLOW DRYING:

If the flow of air is interrupted or stopped, the clothes will not dry, because the moisture is trapped inside the drum. If your complaint is, **SLOW DRYING** of your clothes, then you should check the **VENT** first.

If you have the accordion, or flexible type vent, then check it for kinks or other obstructions. (Figure 20, P. 75) If it is too long, with too many curves or low places where lint and moisture can gather, the best thing you can do is remove it, clean it and cut it to the proper length. If you have the rigid type also check it for proper airflow. Clean it if necessary. To check for proper airflow go to the outside of the house and find the exhaust hood. Feel the air coming out of it. It should be a fairly strong flow of air. If it is weak, then refer to the above. **REMEMBER,** all of the air must pass through the lint screen, so be sure to clean it, **AFTER EVERY LOAD.** And, **NEVER, NEVER** pull the lint screen out when the dryer is running. The lint could be sucked down the chute into the blower. If that happens you may need someone to take it apart and clean it. If your lint screen is located inside the door, you should check the chute (the place where the screen fits) for lint accumulation. If the screen is full of fine lint and won't come off, use a soft brush to clean it. Use water if necessary to aid in cleaning the screen. **REMEMBER, THE BETTER THE AIR FLOW, THE BETTER THE CLOTHES WILL DRY.**

WON'T START:

First make sure the door is closed tightly, re-check the controls and try again. If it still will not start then check to see if you have power. With a gas dryer you can plug a lamp or other device into the outlet to see if it is working. If you don't have any power, check the fuse or circuit breaker. If you don't have a problem there, then call for service.

On an electric dryer you need two fuses or circuit breakers to make it operate. Replace both fuses and try again. If you have circuit breakers then look for a double breaker switch, turn them to **OFF**, re-set them and try again. See the chapter on fuses, P. 76.

Usually, if you have a no-start condition, you have one or more of the following problems; a bad timer, a bad door switch on the newer dryers, an open thermal fuse. A thermal fuse is what prevents the dryer from overheating. It is telling you that you have a plugged vent or other problem.

Before you call for service, check your service manual first for other helpful hints. If you must call for repairs, then clean off the top of the dryer, so the top can be opened if necessary, and make sure there is enough room to move the dryer away from the wall. Above all, provide enough light to see by. While you are thinking about it, **now** would be a good time to update the lighting in your work area. Do it before you need to call the service person. They will be very happy if you do.

ODOR ON CLOTHES:

If the load of clothes, that you just dried, has a very bad odor on them, and you think it smells like gas, do not get excited and think you have a gas leak. Consider this first. Were you or anyone else doing some painting or cleaning with a strong cleaning fluid? Paints, paint thinner, varnish, strong cleaning solutions all have odors that can be sucked into the dryer by the blower. Gas dryers in particular will mix the odors with the open flame, and deposit them in the material. The open flame causes a chemical reaction, and changes it to a different smell completely. To avoid this problem, never run the dryer when any painting or cleaning is taking place. You may have to re-wash the clothes to remove the odor. You may also have to wash out the inside of the drum to rid it of the odor. Use a mild soap and water solution. Most dryers will last from 15 to 30 years if properly cared for. Periodic checks, and not overloading helps extend the life of an appliance. A dry environment also is essential for a long life of an appliance.

COMBINATION WASHER AND DRYER:

If you own a combined unit, that is, a washer/dryer all in one, then you have a special problem.

The blowers on some combination units are unusually weak. The airflow is not as strong as on a free-standing dryer. They require a very short vent to dry properly. If you live in an apartment or a condo, check the length of the vent. If it is too long, the drying time will be extended. Most of the dryers in a condo or apartment are not on an outside wall and the vents are usually longer to reach the outside. Most manufacturers of dryers recommend no more than 20 feet, and 2 elbows in the vent system. If you live in one of these units, and you have to run your dryer for two or three cycles, check the length of the vent first. An unusual amount of lint can accumulate in the dryer when the vent is too long. Most of the combined units use 120 volts, and normally take longer to dry because of the lower ampere rating of the heating coil; and an extra long vent can multiply the drying time.

Dryers 75

HINT: If you want to check to see if the vent is at fault, remove the vent from the dryer, and let the dryer vent into the room while running. It may get a little uncomfortable for a while, but if it gets very hot and humid in the room, you will know the dryer is working and the vent is blocked. If the dryer still doesn't dry, the vent is probably okay and the dryer is defective and needs service. If you must call for service, provide plenty of workspace.

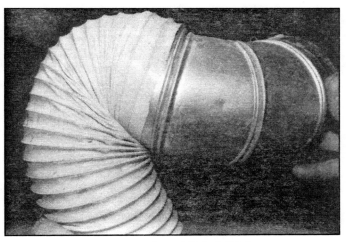

Figure 20 - Vent With A Kink

Chapter 14 FUSES

Any appliance that has a very large cord attached to it and has a three-prong heavy-duty plug uses 220 volts to operate. (Figure 21, P. 80) An electric dryer for instance uses 220 volts. If the appliance runs but won't heat (a dryer as an example) it is often the result of a blown fuse. A 240-volt circuit has two fuses and just one of them can blow by it self. To find out if this is the case, go to the fuse box and look for a black rectangular device about three inches by four inches. (Figure 22, P. 81) This is the fuse holder. It will have a handle in the center. Grasp the handle and pull the holder out. Inside you will find two oblong fuses. They will be marked 35, 45, or 60 ampere. Replace both fuses with the same ampere rating as the ones you remove. Before you plug it back in, go to the appliance and **turn off all controls first.** These types of fuses can look good but in fact can be bad. They can weaken with age and open for no apparent reason. That is why you replace both of them at the same time. Sometimes they are visibly blackened or burned. When you find one like that it usually means a short circuit in the appliance.

Fuses

CAUTION: Make sure all controls are turned off, or the appliance is unplugged before replacing the fuses. Some fuse boxes have the round screw in type. It takes two of them to make a 220-volt circuit. When you locate the right fuses, again, replace both of them.

CIRCUIT BREAKERS:

For 220-volt circuits, look for two levers connected by a handle. This will be a 220-volt circuit. Push the breaker to the **"off"** position and back to **"on"**. That will reset the circuit. In all cases, turn the appliance off before changing fuses or re-setting the breaker. Most appliances use just one fuse or breaker. This is a 120-volt circuit. Look for one bad fuse or breaker that is off. They will not be in the black fuse holder so do not pull it out. That is the wrong circuit for 120 volts.

Most appliances are, or should be, on a separate circuit by themselves. These include the refrigerator, dishwasher, freezer, and washer-dryer. More than one appliance on a circuit causes overloads.

CAUTION: Never remove the cover from the fuse box or the circuit breaker box. You could receive a nasty shock. Leave that to an electrician.

EXTENSION CORDS:

Your household appliances require that no extension cords be used, that they are plugged directly into an outlet with the correct amperage rating. However, if you must use an extension, be sure that a heavy-duty cord is used. Most appliance stores carry cords with the amperage rating to match the appliance.

CAUTION: Never, under any circumstances, use a lightweight cord designed for lamps or small appliances on any appliance discussed in this book. Lightweight cords can overheat and cause a fire. They also can cause low voltage to your appliance, which in turn can burn out a motor or other part.

SCREW IN TYPE FUSES:

If you have a fuse box with the screw-in type of fuses, but some fuses look smaller than others, and are different colors, there is a reason for this. The colored fuses have an adapter screwed into the fuse holder (which is larger), and the colored fuse has a small threaded end. Each color has a different thread

Fuses

and is not interchangeable with each other. The color indicates the amperage of the fuse. BLUE- 15 amp., ORANGE or RED-20 amp., GREEN-30 amp. You cannot put a green fuse in place of a blue fuse. You must use the same color, or amperage, as the one removed. They all have a ceramic thread and are rated as a delay type of fuse. That is, they will sustain a heavy current load for a few seconds before they blow. These are used mainly for air conditioners or refrigerators that have a heavy start up load. There are normal sized fuses that do the same thing, but are not color-coded. They have a metal thread on the base and will fit any standard fuse holder. Delay type fuses all have what looks like a spring inside. This type of fuse requires more caution when replacing them. **NEVER REPLACE A 15 AMP FUSE WITH ONE MARKED 30 AMP.**

Too large of fuse in the wrong circuit can cause the wiring to overheat and cause a fire.

If you purchase an electric dryer, for instance, you want to install the 220 V.A.C. cord, that is the flat looking cord with three prongs on the plug. (Figure 23, P. 81) The middle connector is always the neutral connection and goes to the middle screw on the dryer. The same rule applies to the electric

Secrets of Appliance Repairmen 80

range. The two outer connectors are connected to the outside screws, and it does not matter which one is to the left or the right. Just be sure that the white, or middle wire is on the middle connector. If you have a four wire plug, (used in mobile homes) the green is used for ground, and the white still goes to the middle terminal, and the green to cabinet ground.

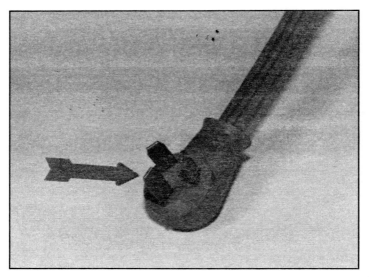

Figure 21 - A 220-Volt Cord For 60-Amp Service

Fuses 81

Figure 22 - 220-Volt Fuse Holder

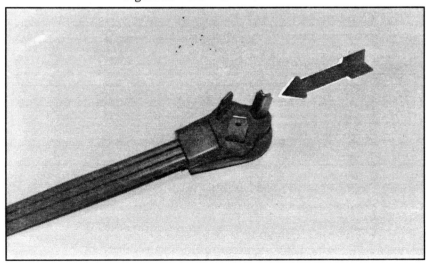

Figure 23 - A 220-Volt Dryer Cord For 30-Amp Service

Chapter 15   HUMIDIFIERS

WHY WE NEED THEM:

During the winter when temperatures get very cold, the atmosphere looses its ability to retain moisture. The colder it gets, the less humidity in the air. The cold air is acting like a giant sponge around our houses, sucking any moisture inside the house to the outside air. The inside air gets so dry that the wood dries out, the furniture creeks, you get sparks from your fingers when you touch a door knob, and your skin and throat can even become dry. We need to replace the lost moisture. This is accomplished with the use of a humidifier.

HOW DO THEY WORK?

The most efficient humidifiers are the ones attached to the furnace. Some merely extend into the plenum over the heat exchanger, and have no moving parts. Instead, they have special plates that absorb water. When the furnace blower comes on, the hot air is passed over the plates and the water is evaporated and returned to the air in the house.

Humidifiers

The best humidifier is attached to the outside of the plenum, and has motors and blowers of it's own. When the furnace blower comes on, the humidifier will also start when the air is dry enough. The hot air is pulled into the humidifier, and is pulled through a wheel holding a special pad that has been moistened from the reservoir below it. The moisture is returned to the airflow, and back to the house. **ALL** humidifiers have a water line attached to them, and all have a float valve to control the amount of water in the reservoir.

If you have a hot water furnace, or electric heat, you cannot use this type of humidifier, because there are no hot air ducts used. Portable units can be used very efficiently as a substitute in any room of the house. They work on the same principle, but use room-temperature air, instead of the hot air from a furnace.

All powered units use a humidistat to control humidity in the air. They are made to sense moisture in the air, and turn the unit off at the proper time. Some controls have a setting for what percent of moisture you want. You merely set it for 40%, 50% or what you are comfortable with. Other controls tell you to set according to the outside temperature and they compensate automatically.

CLEANING:

When air is evaporated, the mineral deposits in the water are left behind. These deposits harden the pads or plates, and leave a residue in the reservoir. The pads or plates should be replaced when they look like a white film is forming, and they feel firm to the touch. Turn the water off, and follow directions in the manual for removing the reservoir. Clean very thoroughly. (Figure 24, P. 85)

Replace the pads and re-assemble. This should be done at the start of the heating season. The portable units require more frequent cleaning, because the water gets stale and will leave an odor in the house. Portable humidifiers do not have a water line attached, and must be filled manually. Use fresh, cold water to fill. When a stale odor is noticed, empty all of the water from the reservoir and scrub clean, following the manufacturers directions for cleaning solutions to be used. Install a new pad every time you clean the unit.

Humidifiers

Figure 24 - Removing The Reservoir

Chapter 16   DEHUMIDIFIERS

Dehumidifiers do the exact opposite of what a humidifier does. Instead of adding moisture to the air, they extract moisture from the air. That is why you should have one in the basement. High humidity causes the water pipes to sweat, and the floor and walls will be damp all the time. In a damp basement mold can form and the air will smell musty.

Dehumidifiers work on the same principal as a refrigerator. They both have a condenser and an evaporator. The evaporator is the part that gets cold. When very humid air is passed over a cold surface, the humidity condenses into water droplets and runs into a collection container. The air is pulled across the evaporator and is pushed out through the condenser at the other end. This is done with a fan, located between the two units. When the air circulates through the dehumidifier, dust is also carried with the air and can eventually accumulate to the point that it stops the airflow completely.

NO WATER FROM DRAIN:

When the airflow stops, the unit will no longer do its job. It must be cleaned. If you want to clean it, first pull the plug from the wall socket.

Second, look on each side, near the bottom of the cabinet for four to six screws. Remove the screws and the cabinet will lift off. Inside you will see the condenser and the evaporator with two fans and two motors, one on each end. Using a scrub brush, carefully clean both units (the condenser and the evaporator) with soap and water. Clean the fan blades at the same time. Look at both ends of the motor for a small hole above the shaft. These are oil holes. Use good quality household oil and put five or six drops in each hole. Wipe up any excess water and replace the cabinet.

HINT: When using your humidifier, be sure that you have all the windows closed, and the door to the basement closed also.

If the doors and windows are open, then you might as well set it in the back yard and try to dry up the world. They are designed for a closed environment, and an open window lets the humidity in faster than it can be removed.

CAUTION: Check the floor drains from time to time. Every basement should have a floor drain with a metal grate cover. Below the grate in the drain a trap has been installed to trap sewer gas. The water in the trap can dry up and let sewer gas seep in, giving off a bad odor. Pour a quart of water in the drain once a week to prevent that from happening. Some brands have a drain in the collection pan that allows a hose to be attached. Run the other end of the hose to a floor drain.

The water draining from the hose will keep the trap full for you, but if you have more than one drain check the others on a regular basis.

Chapter 17   HOT WATER TANKS

When you think of water heaters, you are thinking of the most dependable and least understood appliance in your house. Most commonly they are called **HOT** water heaters. This is incorrect. They do not heat <u>hot</u> water they heat cold water. Your heater will give you years of carefree service with very little maintenance required.

HOW TO FLUSH THE TANK:

The most important preventive maintenance function needed is to flush the tank out once a year. Most people either do not flush it out, or they do it wrong. The proper way to flush it out is to use your garden hose. First, find the drain on the tank. It looks the same as the faucet on the outside of the house, where the garden hose screws on. They are almost identical to each other. Screw the hose to the drain on the tank and put the other end of the hose into the utility sink, making sure it is secure. Us a weight of some kind if you have to. If you do not have a utility sink, then find the standpipe, where the washer drain hose is installed. Remove the drain hose and put the end of the garden hose into the pipe at least a foot. When the hose is secured properly, return to the water tank and turn the faucet on.

Run the water slowly at first, but increase the flow until you have full flow and pressure. Keep the flow going at least three minutes but no more than five minutes. If the sink or the standpipe won't take the amount of water required, you may have to run the hose outside to the curb or driveway. If you must run it outside, remember, the hot water can kill flowers and shrubs and maybe even the grass, so run it to a safe place. Inside the tank there is a tube that takes the incoming cold water, and directs it to the very bottom of the tank.

Hot water raises naturally, but the cold water entering the bottom forces the hot water to the top, and out the pipe to your sink. The pressure stirs up any sediment lying on the bottom but is not strong enough to remove it from the tank. When you flush the tank, all the force stays at the bottom and the sediment is forced out the drain. If the tank is not flushed on a regular basis the sediment builds up and causes rust. A gas-fired tank will leak sooner, because the build up in the bottom traps the heat and causes the metal bottom to eventually warp, and start leaking.

RUNNING OUT OF HOT WATER:

If you run out of hot water in a matter of minutes, it doesn't necessarily mean the tank needs replacing. The fault may lie in the tube inside the tank. This was described above.

If the tube (commonly called a dip tube) develops a hole or crack, or breaks off entirely, near the top, the cold water will mix with the hot water at the point of the leak instead of from the bottom of the tank, cooling the hot water very rapidly. The dip tube can be replaced by any plumber and save you the cost of a new tank.

If you must replace the tank because of a leak, then consider what size is best suited for the size of the family, and the size of the house. That is, do you have more than one bathroom and shower or tub? If you have a large family, you may want a larger capacity tank with a fast recovery rate. You could have a shower, a dishwasher and a washing machine all operating at the same time. Check with a dealer for recovery rates for gas and electric models to see which one is best suited for your needs. There are water heaters available that do not store hot water at all. They only heat what is needed when the faucet is turned on. This type has a very high B.T.U. rating. The B.T.U. (British Thermal Unit) tells you how fast an appliance will heat up. A 60,000 B.T.U. will heat faster than 30,000. That is, they use a lot of gas at any given time and may be too expensive to operate for a large family. Granted, if you are gone for a month on vacation, no gas is used while you are gone.

If there are only two people in the house, you may want to buy a smaller unit to meet your needs. Check around before you buy to get the best bargain, but be sure what you purchase meets your needs.

Chapter 18   MISCELLANEOUS

FLUORESCENT LIGHTS:

Fluorescent lights are trouble free for the most part, but replacing the tubes can be tricky. First, determine if a new tube is needed. If the fixture has one tube in it, and it is flashing of and on, the problem may be the starter. Some fixtures have the starter on the end of the fixture where it is easy to change. Others are located behind, or above the tube. The tube must be removed to gain access to the starter. Once you have located the starter, (Figure 25, P. 96) grasp it with your thumb and forefinger and turn counter clockwise ¼ turn. It should pull out easily then. Take note of the letters and numbers on the starter. It should read, as an example, FS 25, or FS 15, be sure to replace it with the same rating numbers. If the tube is blackened at the end, inside the glass, it means the tube must be replaced. It's a good idea to replace both the starter and the tube at the same time.

CAUTION: All fluorescent tubes have two pins on each end of the tube. These are the electrical contacts that make the tube light. When removing the tube from the fixture, it must be turned

¼ turn to release the pins. When the pins are released, the tube should pull straight out of the holders. When installing a new tube take care not to bend the pins. Reverse the procedure to install the new part.

A two light fixture rarely uses a starter. This type of light uses a ballast instead. If one tube is dark and the other is very dim, it means the dark one is burned out and needs to be replaced. Both tubes must be good in order to light. When the tube is black inside it should be replaced. Remember, it takes two good tubes to work and will not light with just one.

A four-tube fixture is the same as a two tube except it uses two ballast transformers instead of one. These also work in pairs. If you replace the tubes and they still do not light, the ballast may be at fault. It's best if you do not attempt to replace the ballast. Check the yellow pages first for small appliance repair shops. If your two light fixtures have two starters, replace them both at the same time.

ELECTRIC CAN OPENERS:

This appliance has very little maintenance problems, but you should be aware of a simple procedure that can be done to help it work better. Some openers have a round, sharp cutting wheel that actually opens the can. Others have a knife-like device that cuts the can open. If you look at the center of the device you will see the head of a screw. Remove the screw and the cutter will come off to be cleaned. This should be done often because every time a can is opened, food is deposited on the surface of the cutter. Wash the rest of the opener at the same time and be sure to wipe off the round part that looks like a gear. This is the device that turns the can so the cutter can work, and is not removable. If the cutter is worn and seems to wobble when used, it should be replaced.

If you still have the book that came with the opener, then check to see if the manufacturer offers replacement parts. Often they will list a part number for parts that can be replaced and give an address of where to write to, and perhaps a phone number also.

TOASTERS:

Toasters last for years and are virtually trouble free, but cleaning is a must. An accumulation of breadcrumbs could overheat and cause a fire. Unplug the toaster and hold it over a paper bag or wastebasket while removing the panel on the bottom side of the toaster. There is usually a lever that turns ¼ turn to release the panel. As the panel comes off, the crumbs will fall out on their own. You may have to shake the toaster a bit to loosen any leftover crumbs. If everything doesn't come out, never use a knife or other utensil to scrape with. If you must use anything, use a very soft brush, such as a clean paintbrush with long bristles or a pastry brush. Do not move any wires or touch

Miscellaneous 97

the heating element (the flat, bare wire) or attempt any adjustments. When you are done cleaning, replace the panel the same way you removed it.

CAUTION: If your toaster does not heat up, do not try to repair it. Look in the yellow pages under small appliance repair shops and call first to see if they will repair toasters.

CABARET FANS:

Overhead fans should be cleaned at least once a year. Use a good household cleaner and wash the blades, top and bottom. Do not splash any water into the motor housing. After the blades are dry, use a good quality wax on them. A clean blade will push twice as much air as a dirty one.

When installing a new unit follow directions very carefully when hooking up the wires. If you have a three-speed fan all you need to do is attach the black wire to the black, white to white and green to ground.

Never install a speed control in place of the switch on the wall. A speed control will burn out the motor. The pull switch on the fan controls the speed. Also, never start the fan on low speed. Start it on high (usually the first pull of the string) and then switch it to low. With these precautions in mind the fan should last for years. If you do have a problem there are shops that specialize in overhead fan repairs.

GARAGE DOOR OPENER:

If the opener fails to activate when the remote control button is pushed you should do two things. First, make sure you have power to the opener and second, replace the battery in the remote control. It should work if the power is on and the battery is new. If not, you may have to get a stepladder and check the unit itself. Look on the drive motor for a small red button. This will be the overload switch. Pull the plug from the power source and push the red button in as far as it will go. If it stays in, you have found the trouble. Now, plug the unit back in (KEEP HANDS AWAY FROM GEARS OR CHAINS) and see if it will run. If the re-set button pops out again that means the door is binding or the controls are out of adjustment.

DO NOT ATTEMPT TO ADJUST UP OR DOWN STOPS. If it keeps going up and down or goes half way and shuts off, leave it alone. Call for a repairman.

MAINTENANCE:

You should apply grease and oil to the proper places at least once a year. Put 4 or 5 drops of good quality oil on each of the door rollers using a zoom spout oiler (See Figure 26, P. 100). On the chain or worm gear drive use good grease like Lubriplate (Figure 27, P. 100).

ADJUSTMENTS:

The only adjustment you should try to make is the reverse control. This is usually called "FORCE". Lay a cardboard box directly under the door. Now activate the door so that it comes down on the box. It should reverse without crushing the box. If it crushes the box, set the force control towards "DECREASE". Adjust until the box is not crushed under the door.

Secrets of Appliance Repairmen 100

Figure 26 - Oil Here

Miscellaneous 101

Figure 28 - For Cleaning Lint On This Dryer

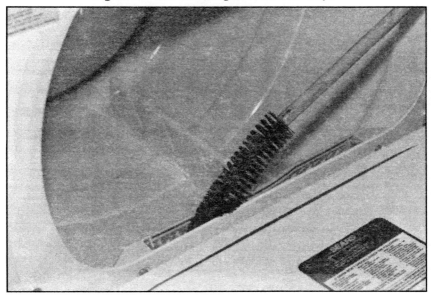

Figure 29 - Or This Dryer

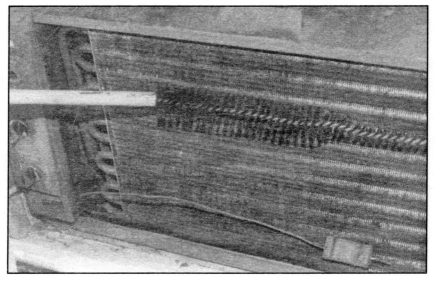

Figure 30 - Use On Air Conditioning Condenser

Chapter 19 — A GLOSSARY OF TERMS USED

AIR GAP: A device used on built-in dishwashers that prevents the drain hose from coming in contact with the house drain. Also prevents water from the sink backing up into the dishwasher. Required in some cities by law.

CONDENSER: A condenser is part of every refrigeration device made. It is the part that the hot gas is pumped to for cooling. When working correctly, hot air will be felt coming out of it. It will be hot to the touch. Gas is converted back into liquid here.

EVAPORATOR: An evaporator is also part of every refrigeration device. The evaporator is very cold. The liquid (freon) is converted into vapor in the condenser.

FOAMING: Foaming in a dishwasher is the same as in the washing machine when detergent is added. Foam is the suds floating on top of the water and is unwanted in a dishwasher.

FREON: The commercial name given a refrigerant used in cooling and freezing devices. One chemical is called dichlorodiflouromethane.

FUSE BOX: A device that distributes all of the electrical circuits to different rooms in the house. Each circuit has it's own fuse. Same as CIRCUIT BREAKER BOX.

HUMIDISTAT: A device that works the same as a thermostat, except that it measures humidity instead of heat. It turns the dehumidifier on and off according to the moisture present in the air.

IMPELLER: A device used in pumps to move water at a rapid rate. It is round and has blades that push the water into an outlet on the pump and is directed to its destination. Fastened directly to a motor shaft, and if broken will not pump water.

KINK: A kink is what happens when you bend the garden hose to stop the water at the open end. A kink will stop water from

A Glossary of Terms Used

flowing through a hose, and will prevent air from flowing through a dryer vent.

MULLION HEATER: This is a device mounted around the door opening of the refrigerator to keep the metal warm. Mounted behind the plastic liner, it stops sweating in hot, humid weather.

ORFICE: An orfice (sometimes spelled orifice) is a regulator of gas flow. Regulates the amount of gas going to a pilot light or gas burner. Comes in different sizes according to needs. Usually has a very small opening.

OVERLOAD: A device that protects electric motors from overheating. The overload will stop the flow of electricity to the motor if it can't start. Some will re-set automatically when the motor cools. Others must be re-set and have a button to push manually.

PLENUM: A plenum is the sheet metal part that sits on top of the furnace and collects the hot air. The ducts that go to the separate rooms of the house are attached to the plenum.

RE-SET BUTTON: See above.

ROTOR: The part of a motor that has a shaft that turns. Will have an impeller, a fan or a blade attached to the shaft.

SIPHON: Any container of liquid with a hose attached can be siphoned. Even if the hose is higher than the liquid it will siphon when the END of the hose is LOWER than the liquid and the end on the container is below the liquid level. All that is required is to start a small flow of liquid, and it will continue to flow.

THERMAL FUSE: A thermal fuse is like a thermostat. It shuts off at a predetermined temperature, but unlike a thermostat, it will not re-set. It is a safety that prevents overheating and must be replaced. If a vent is blocked, the thermal fuse may open.